My Book

This book belongs to

Cover Design by :
Gowri Vemuri

First Edition :
January , 2020

Author :
Gowri Vemuri

Edited by :
Raksha Pothapragada
Ritvik Pothapragada

Questions: mathknots.help@gmail.com

NOTE : VDOE is neither affiliated nor sponsors or endorses this product.

Dedication

This book is dedicated to:

My Mom, who is my best critic, guide and supporter.

To what I am today, and what I am going to become tomorrow,

is all because of your blessings, unconditional affection and support.

This book is dedicated to the

strongest women of my life ,

my dearest mom

and

to all those moms in this universe.

G.V.

The Virginia Board of Education and Virginia Department of Education (VDOE) have developed the Virginia Assessment Program (VAP) to measure and evaluate students' academic progress in the Standards of Learning (SOLs). The SOLs indicate Virginia's expectations for what students should know and be able to do in the subject areas of reading, writing, mathematics, science, and history/social science.

Students in grades 3-12 to take the Standards of Learning (SOL) assessments each year. Some of the tests are required by all students each year, and others are required only at specific grade levels. Additionally, with the removal of some SOL tests in recent years, the VDOE assigned the responsibility of the creation and administration of alternate, performance - based assessments on local divisions. Student scores from these tests determine a school's and the division's state accreditation and measures progress toward meeting federal targets.

Virginia Standards of Learning (SOL) tests are generally given online unless a student has an identified and documented need to be assessed using paper,pencil format. The test question format is typically multiple choice, and each test contains some technology enhanced items.

GRADE 3-8 STANDARDS OF LEARNING(SOL) TESTS

GRADE 3	GRADE 4	GRADE 5	GRADE 6	GRADE 7	GRADE 8
	VIRGINIA STUDIES				WRITING
MATH	MATH	MATH	MATH	MATH	MATH
READING	READING	READING	READING	READING	READING
		SCIENCE			SCIENCE

NOTE : VDOE is neither affiliated nor sponsors or endorses this product.

END OF COURSE STANDARDS OF LEARNING (SOL) TESTS

GRADE 9	GRADE 10	GRADE 11
ALGEBRA I	GEOMETRY	ALGEBRA II
EARTH SCIENCE	BIOLOGY	CHEMISTRY
WORLD HISTORY I	WORLD HISTORY II	VIRGINIA & US HSTORY
		WORLD GEOGRAPHY
		ENGLISH : READING
		ENGLISH : WRITING

Any Student taking one of the courses listed here is expected to take the corresponding end-of-course SOL test. The grade levels depicted here represent grade level at which students typically participate in these courses.

NOTE : VDOE is neither affiliated nor sponsors or endorses this product.

SOL Test Scoring and Performance Reports:

Standards of Learning assessments in English reading, mathematics, science and history/social science are made up of 35-50 items or questions that measure content knowledge, scientific and mathematical processes, reasoning and critical thinking skills. English writing skills are measured with a two-part assessment that includes multiple-choice items and an essay.

Student performance is graded on a scale of 0-600 with 400 representing the minimum level of acceptable proficiency and 500 representing advanced proficiency. On English reading and mathematics tests, the Board of Education has defined three levels of student achievement: basic, proficient, and advanced, with basic describing progress towards proficiency.

Performance Achievement Levels:

- The achievement levels for grades 3-8 reading and mathematics tests are: *Pass/Advanced*, *Pass/Proficient*, *Fail/Basic*, and *Fail/Below Basic*.
- The achievement levels for science tests, history tests, and End-of-Course (EOC) tests* are: *Pass/Advanced*, *Pass/Proficient*, and *Fail/Does Not Meet*.
- The EOC Writing (2010 SOL) test, EOC Reading (2010 SOL) test, and EOC Algebra II (2009 SOL) test have an achievement level of *Advanced/College Path* in place of the *Pass/Advanced* achievement level.

NOTE : VDOE is neither affiliated nor sponsors or endorses this product.

INDEX

FORMULA SHEET

1. Area of a triangle

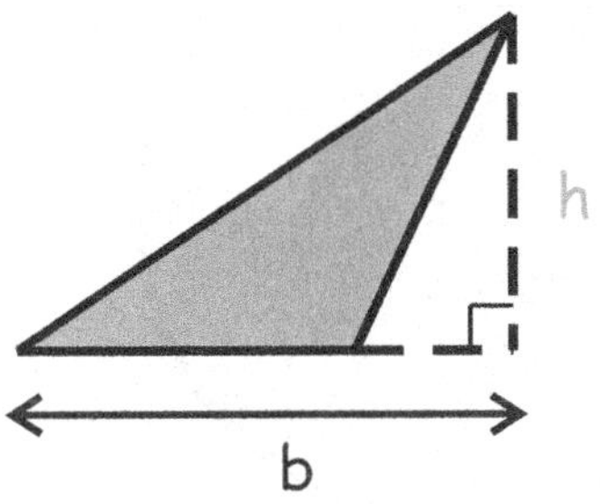

$A = \frac{1}{2} bh$

2. Area of a parellelogram

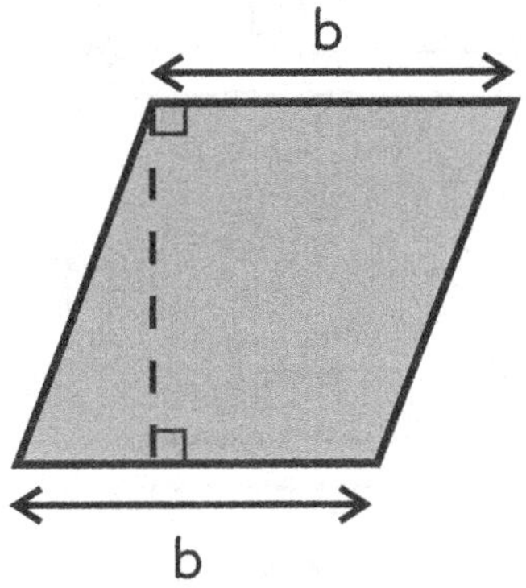

$A = bh$

3. Volume and Surface area of a cuboid

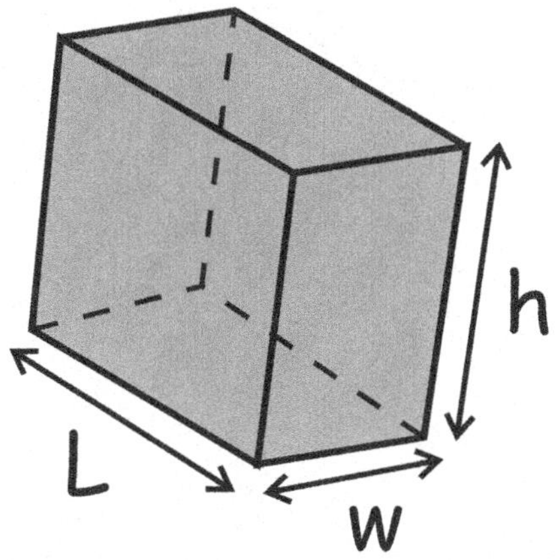

$V = lwh$

$S.A = 2(lw + lh + wh)$

4. Perimeter and Area of a Square

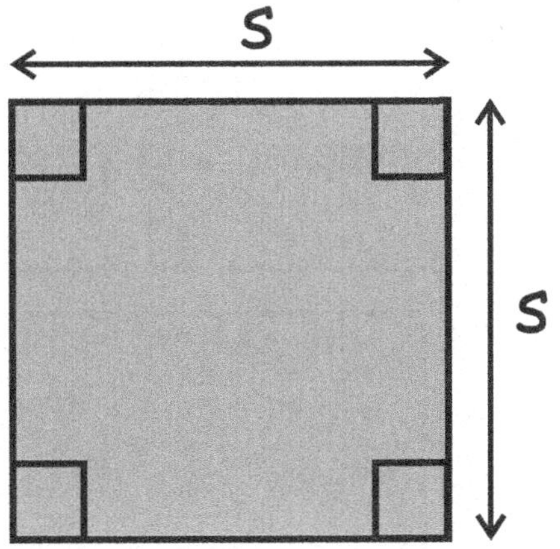

$P = 4s$

$A = s^2$

5. Area of a Trapezium

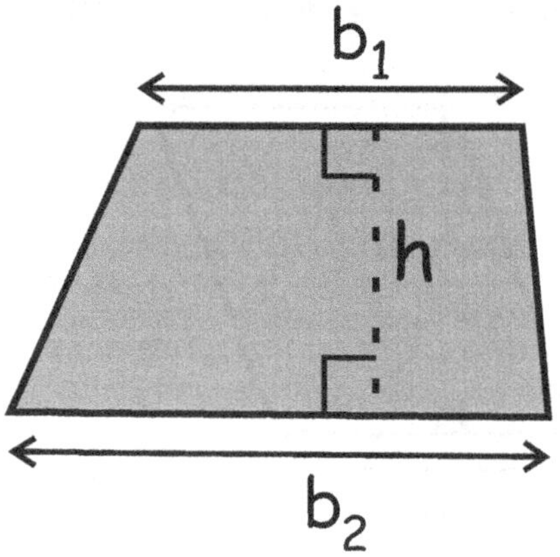

$$A = \frac{1}{2}h(b_1 + b_2)$$

6. Circumference and Area of a Circle

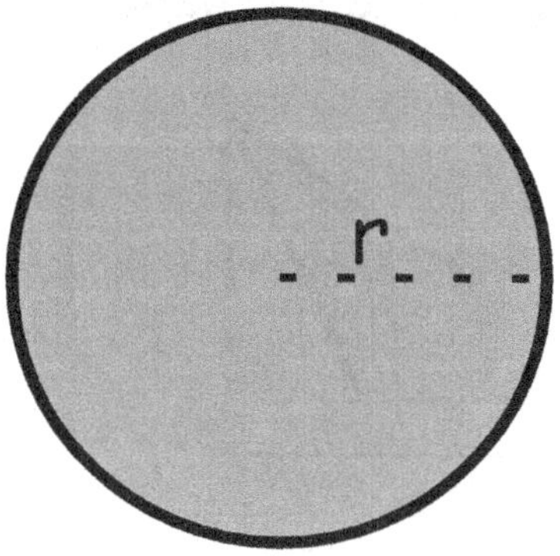

$$c = 2\Pi r$$
$$A = \Pi r^2$$

pi

$$\Pi = 3.14$$
$$\Pi = \frac{22}{7}$$

7. Perimeter and Area of a Rectangle

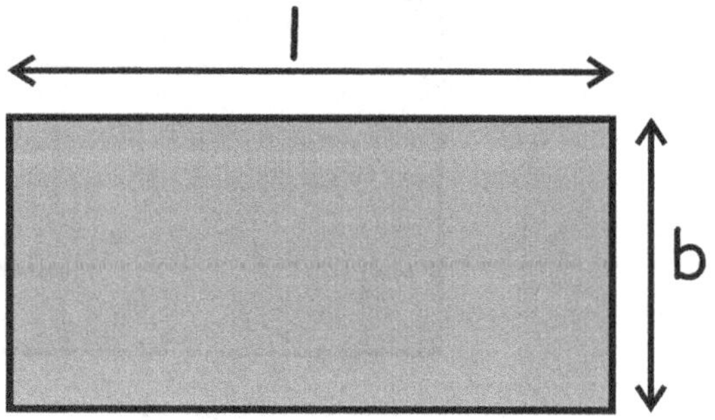

Area = l b ×
Perimeter = 2(l + b)

8. Acute angle

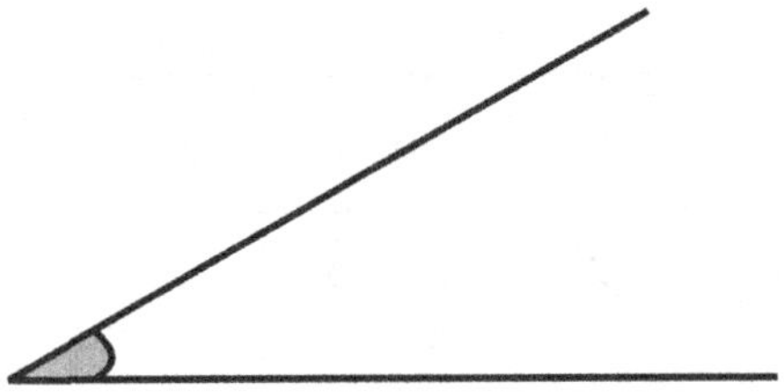

Any angle less than 90^0
is called as acute angled

9. Right angle

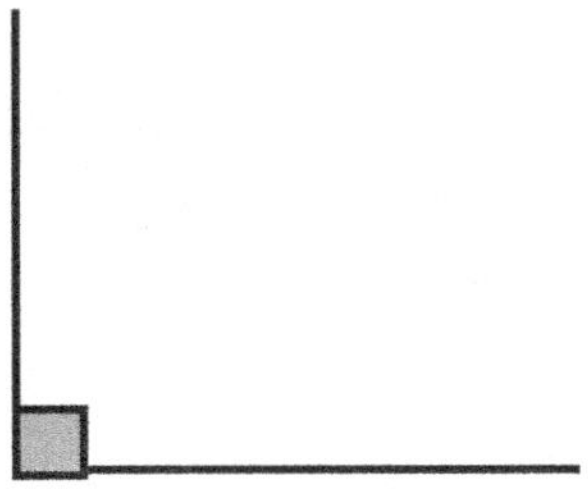

An angle equal to 90^0
is called as right angled

10. Obtuse angle

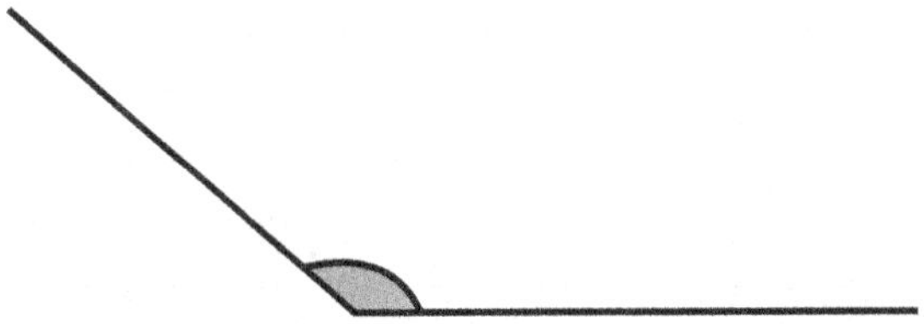

Any angle greater than 90^0
is called as obtuse angled

11. Perimeter of a polygon KLMNOP

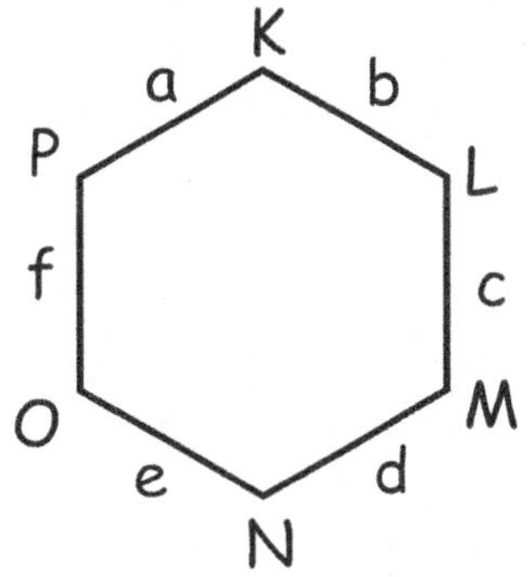

Perimeter =
(a + b + c + d + e + f) units

Abbreviations

milligram	mg
gram	g
kilogram	kg
milliliter	mL
liter	L
kiloliter	kL
millimeter	mm
centimeter	cm
meter	m
kilometer	km
square centimeter	cm^2
cubic centimeter	cm^3

year	yr
month	mon
hour	hr
minute	min
second	sec

volume	V
total Square Area	S.A
area of base	B
ounce	oz
pound	lb
quart	qt
gallon	gal.
inches	in.
foot	ft
yard	yd
mile	mi.
square inch	sq in.
square foot	sq ft
cubic inch	cu in.
cubic foot	cu ft

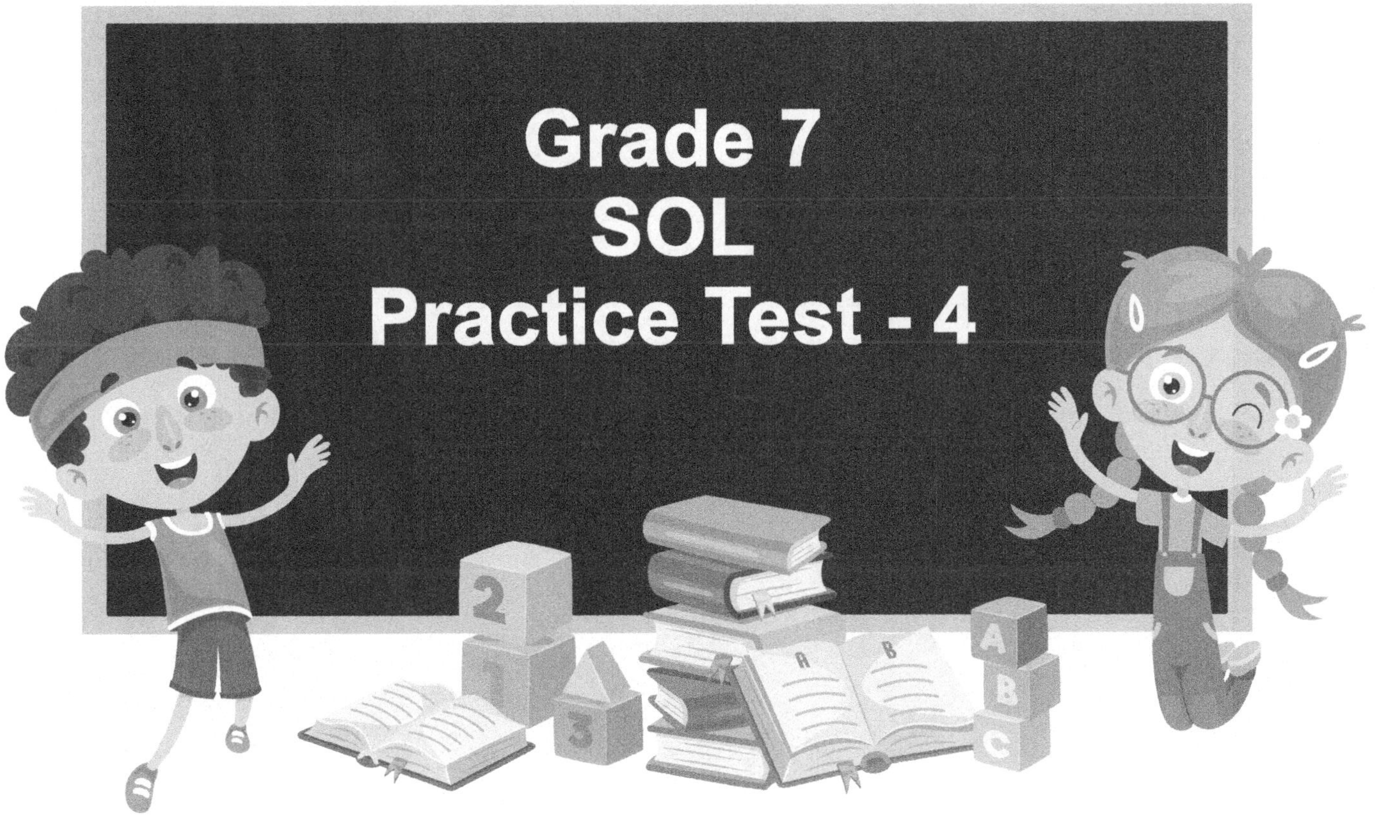
Grade 7
SOL
Practice Test - 4

1. Simplify

$$\frac{(-12)}{((-11) - (-6) + 4)}$$

(A) 23

(B) 12

(C) -2

(D) 6

2. Simplify

$$(11) - (-13)$$

(A) 14

(B) 20

(C) 24

(D) 16

3. Express the given number in scientific notation

409000

(A) 40.9×10^{6}

(B) 40.9×10^{5}

(C) 40.9×10^{-6}

(D) 4.09×10^{5}

4. A painting is 6 in wide and 2 in tall. If it is enlarged to a height of 8 in, then how wide will it be ?

(A) 24 in

(B) 26 in

(C) 2 in

(D) 32 in

5. Simplify

$$(-9) * (-3) - 8 * 4$$

(A) -10 (B) 15

(C) -5 (D) 3

6. Which of the below is an example of "Rotation 90° counterclockwise about the origin" ?

(A)

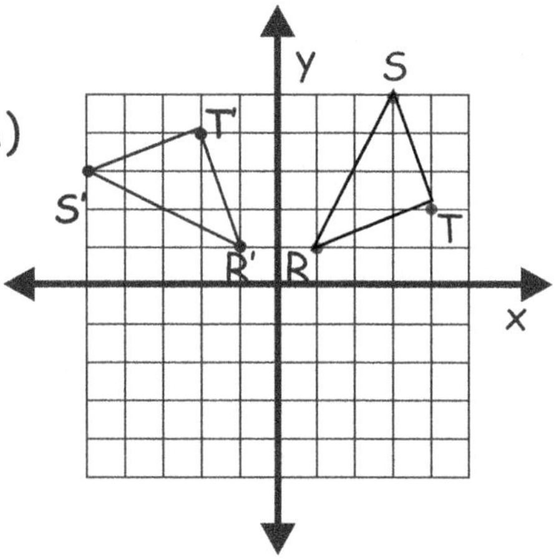

(B)

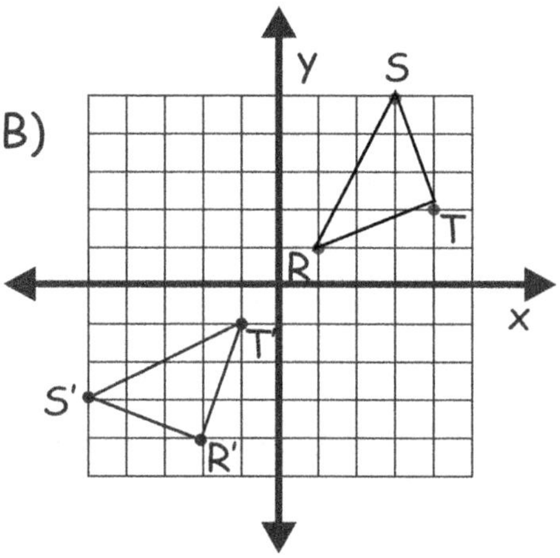

(C)

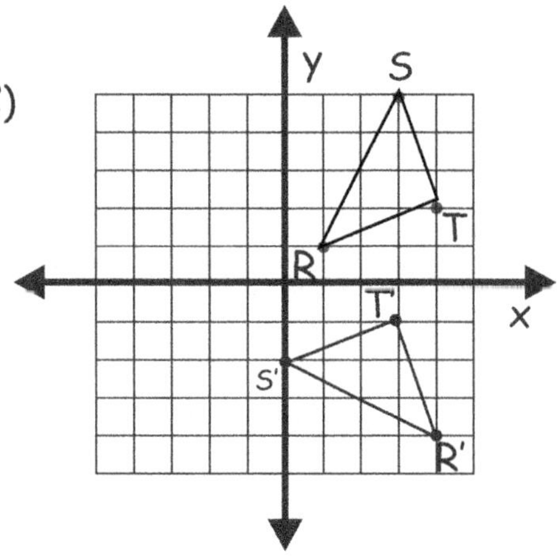

(D)

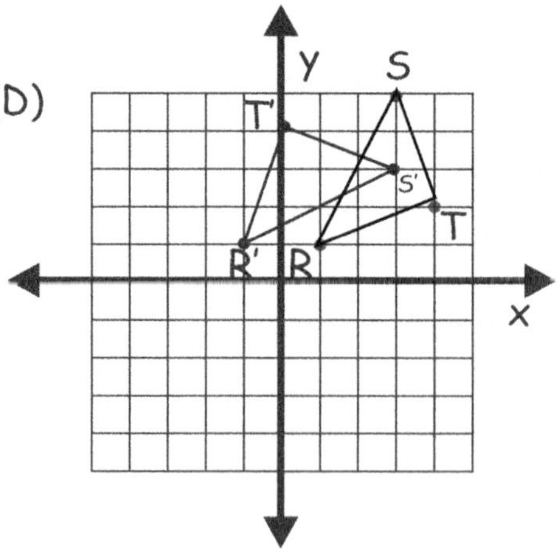

7. Determine the quadrant of the below point belongs to.

(-7, 9)

(A) Quadrant I (B) Quadrant II

(C) Quadrant III (D) Quadrant IV

8. Find the volume of the cylinder with a diameter of 8 in and a height of 6 in.

(A) 301.6 in^3 (B) 75.4 in^3

(C) 66.4 in^3 (D) 150.8 in^3

9. Find the area of the given quadrilateral.

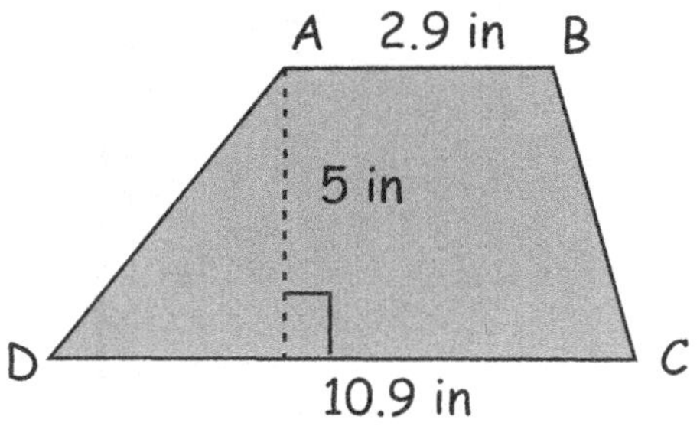

(A) 17.3 in^2 (B) 38.3 in^2

(C) 76.6 in^2 (D) 34.5 in^2

10. Find the total surface area of the given three dimensional shape.

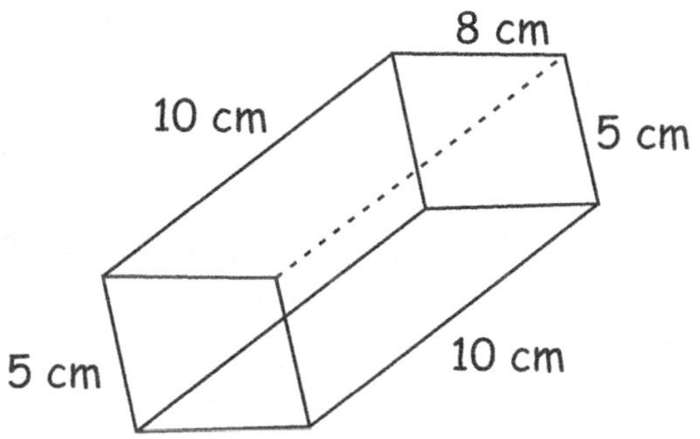

(A) 397.8 cm^2 (B) 294.1 cm^2

(C) 350.1 cm^2 (D) 340 cm^2

11. In how many ways a captain and a co-captain can be chosen from a team of 14 field hockey players ?

(A) 182

(B) 180

(C) 82

(D) 90

12. Which of the below scatter plots can be plotted using the given data ?

X	10	32	32	48	60	60	60	62	63	80
Y	6	2.8	3.6	8	6.5	0.7	10	5.2	7.6	5

(A)

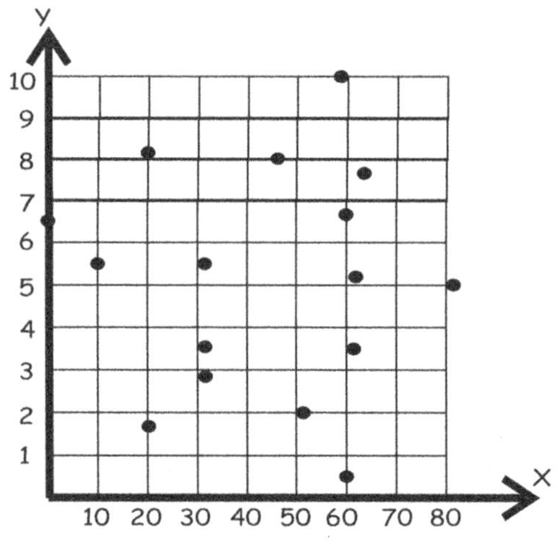

(B)

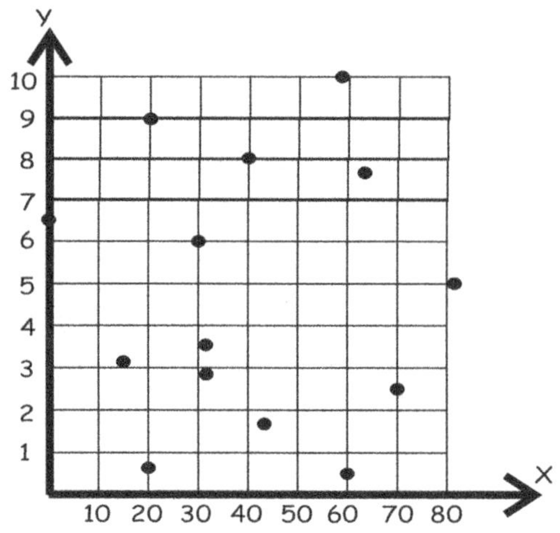

(C)

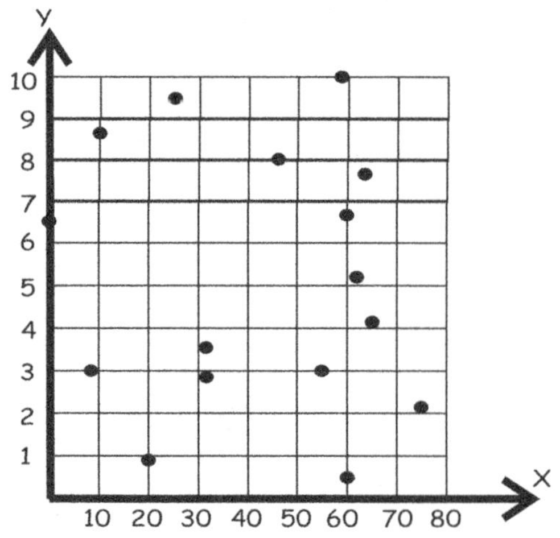

(D)

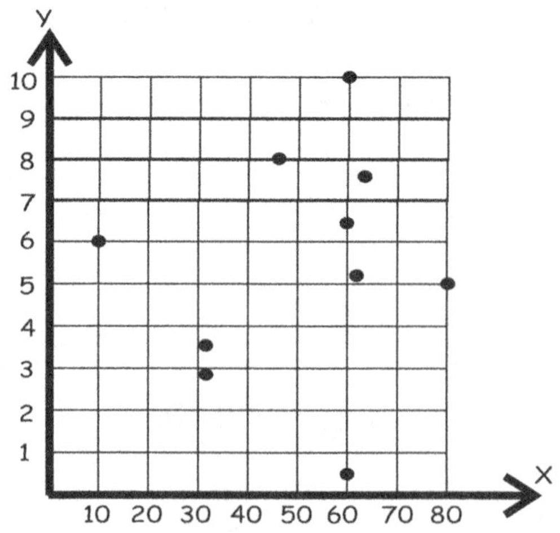

13. Find the median for the given below data.
30.4 , 23.6 , 34 , 40 , 56.6 , 63.2 , 56.4 , 68.2 , 52.6 , 50.6 , 35.2

(A) 50.6 (B) 52.6

(C) 40 (D) 30.4

14. Find the average for the following data.

Inches	Frequency
64	2
65	3
66	1
67	2
69	5
70	2
71	2
72	3
73	1
74	1

(A) 65.77 (B) 67.87

(C) 68.77 (D) 66.50

15. Simplify $-168 = -12b$

(A) -10 (B) 15

(C) 7 (D) 14

16. Translate the verbal expression to an algebraic expression.

13 to the x is less than 31.

(A) $13 + x < 31$

(B) $13^x < 31$

(C) $x^{13} < 31$

(D) $13 - x < 31$

17. Simplify

$(-22) + (-16)$

(A) -29

(B) -56

(C) -39

(D) -38

18. Which of the following is the correct statement for the given expression ?

10^t

(A) t to the 10th

(B) 10 to the t

(C) t less than 10

(D) t more than 10

19. Given below is an expression.
What symbol shall be placed in the below box ?

18x + 97 [?] 20

(A) =

(B) +

(C) >=

(D) None

20. Translate the verbal expression to an algebraic expression.

6 less than n is less than or equal to 6

(A) $n - 6 \leq 6$

(B) $n^2 \leq 6$

(C) $6^n \geq 6$

(D) $6^2 \leq 6$

21. Simplify

$27 = x + 10$

(A) 37

(B) 11

(C) 17

(D) $2\frac{7}{10}$

22. Which of the below options describes $2g - 81 < 19$?

(A) Equation

(B) Expression

(C) Inequality

(D) Less than

23. Which of the below is the geometric series ?

(A) 3, 9, 18, 24, 36

(B) 1, 9, 27, 81

(C) 3, 9, 27, 54

(D) 3, 9, 27, 81

24. Larry ate 15 sugar cookies made by his mom. He ate $\frac{3}{10}$ of the cookies made by his mom. How many cookies are left ?

(A) 4.5 (B) 35

(C) 58 (D) 50

25. Ava has $5 savings. She wants to buy a bracelet and her savings cover $\frac{1}{4}$ of its cost. How much did the bracelet cost ?

(A) $30 (B) $25

(C) $15 (D) $20

26. Write an equation showing the relation between x & y.

x	16	18	20	40
y	28	30	32	52

(A) $y = x + 12$ (B) $y = x + 2$

(C) $y = x + 20$ (D) $y = x - 12$

27. The table below shows Math test scores of Grade 6 student

78 , 79 , 70 , 77 , 69

The teacher missed to add a student score. If the median and mode of the data is same. What is the missing score ?

(A) 77 (B) 74

(C) 79 (D) 70

28. A dress is priced at $134.99 and the store is offering a 40% discount. What is the selling price of the dress ?

(A) $188.99

(B) $114.74

(C) $80.99

(D) $121.49

29. One box of tomatoes cost $3. How many boxes can be purchased with $21 ?

(A) 7

(B) 63

(C) 6

(D) 5

30. A micron is a unit used to measure specimens viewed with a microscope. One micron is equivalent to 0.00003937 inch. How is this number expressed in scientific notation ?

(A) 393.7×10^{-7}

(B) 3.937×10^{5}

(C) 3937×10^{-8}

(D) 3937×10^{8}

31. Find the equivalent expression to the below:

$$5.4(6.78 + 3.3)$$

(A) 5.4 * 6.78 + 5.4 * 3.3

(B) 5.4 * 6.78 * 5.4 * 3.3

(C) (5.4 + 6.78) * (5.4 * 3.3)

(D) 5.4 + 6.78 + 5.4 + 3.3

32. Which of the below is less than 125% ?

(A) $\frac{1}{80}$

(B) 540

(C) 125

(D) $1\frac{1}{4}$

33. Quadrilateral ABCD and EFGH are congruent. Then what is the ratio of length $\overline{BC}$ to length $\overline{FG}$?

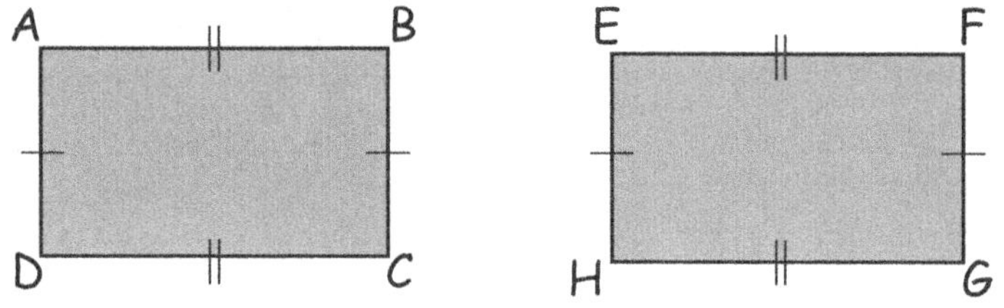

(A) 1 : 2

(B) 1 : 1

(C) 2 : 3

(D) 1 : 4

34. Polygon below is named as ______________

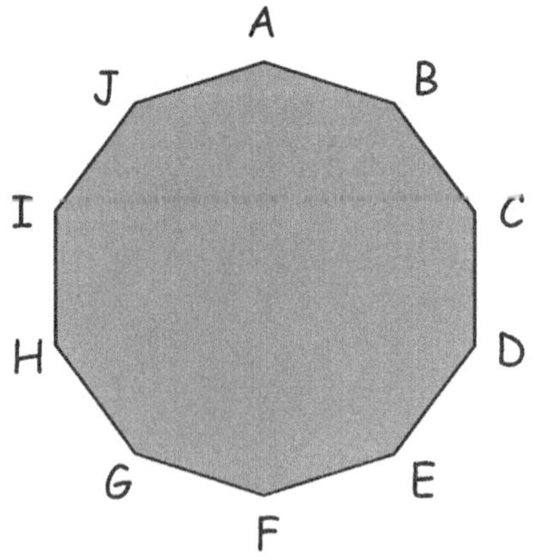

(A) Nanogon

(B) Decagon

(C) Quadrilateral

(D) None

35. Find the value of x from the below triangles.

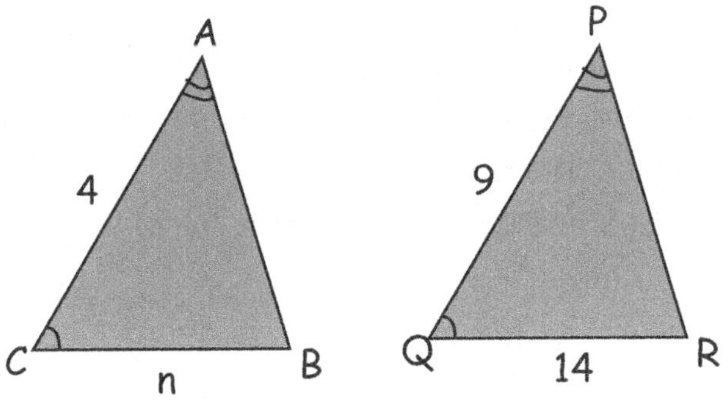

(A) $14n = 36$

(B) $\frac{n}{4} = \frac{9}{14}$

(C) $\frac{4}{n} = \frac{14}{9}$

(D) $\frac{n}{4} = \frac{14}{9}$

36. A submarine is 2340 meters below the sea level. Later it ascended 650 meters, but then descended another 300 meters and finally stopped at this location. Where was the submarine in relation to sea level when it stopped ?

(A) -1990

(B) -3290

(C) 3290

(D) 1900

37. Below equation is an example of

$$(665 + 335) + 445 = 335 + (665 + 445)$$

(A) Identity Property of Multiplication
(B) Distributive Property
(C) Associative Property of Addition
(D) Commutative Property of Multiplication

38. Which of the below is the property of a Parallelogram ?

(A) Opposite sides are not parallel and equal to each other
(B) Opposite angles are not equal
(C) Diagonals bisect each other
(D) Opposite sides are not equal in length

39. Which of the below is the property of Trapezoid ?

(A) Two pair of opposite sides is parallel
(B) Diagonal intersect each other in the same ratio
(C) Two adjacent angles are supplementary
(D) Opposite sides are of equal in length

40. Which polygon has 2 times as many angles as a triangle ?

(A) Octagon
(B) Pentagon
(C) Square
(D) Hexagon

41. Express the given fraction as a percentage.

$$\frac{14}{25}$$

(A) 56%
(B) 14.25%
(C) 0.56%
(D) 560%

42. Express the given fraction as a percentage.

$$\frac{15}{16}$$

(A) 0.9375%
(B) 153.75%
(C) 9375%
(D) 93.75%

43. A local pizza store donates 10% of its sales of every Saturday. Based on the information given

(A) Total sales amount of last Friday was $880 and donated $44
(B) Total sales amount of last Friday was $350 and donated $3500
(C) Total sales amount of last Friday was $65 and donated $650
(D) Total sales amount of last Friday was $3500 and donated $350

A company manufactures a number of skate boards over the years as shown in the below graph. Answer the questions 44 - 48. (Units in millions)

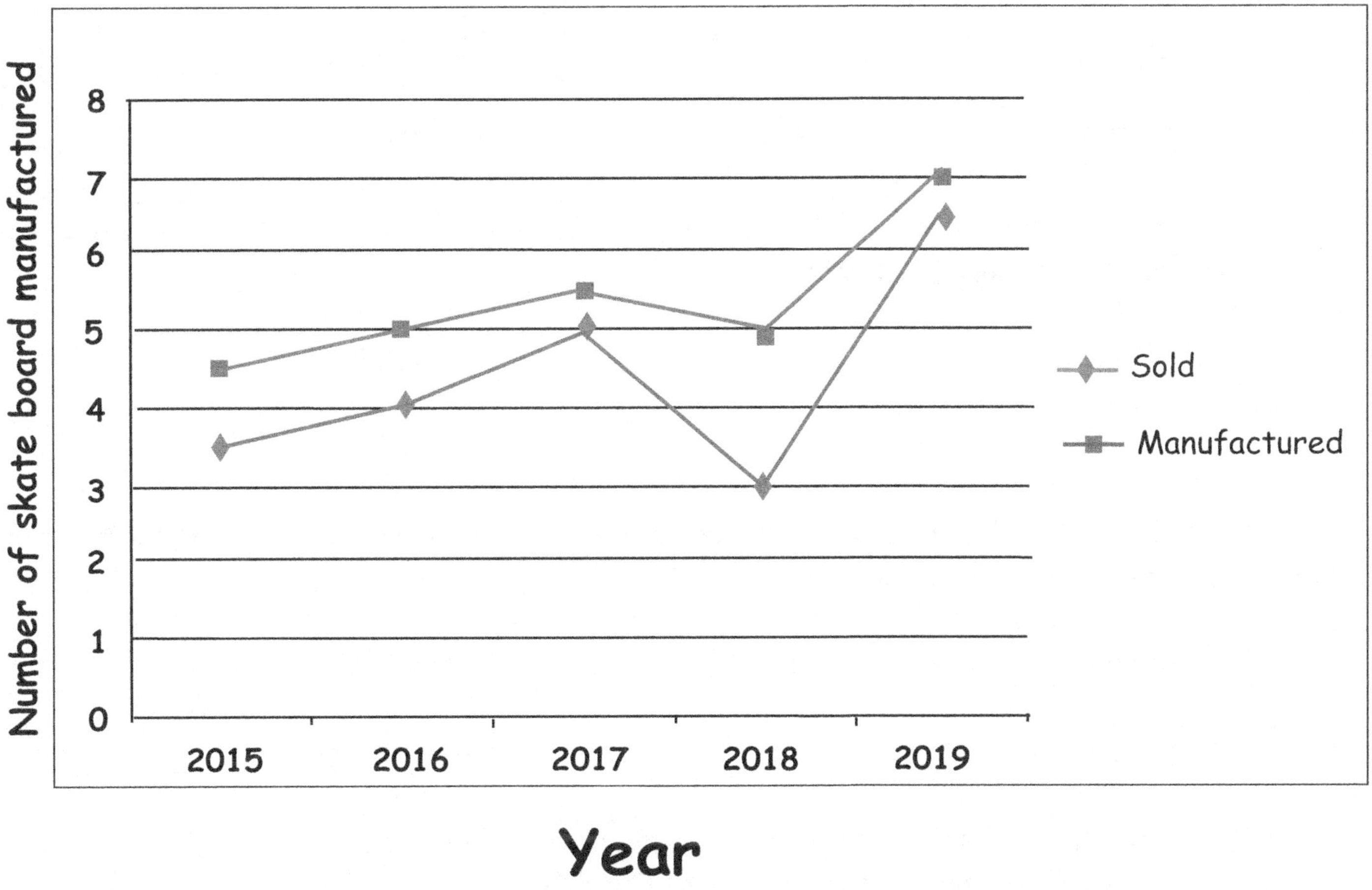

44. What is the average number of units sold over the years ? (In millions)

(A) 44
(B) 4.4
(C) 0.44
(D) 440

45. Find the ratio of the difference between the number of units manufactured and sold in the year 2017 to the difference between the number of units manufactured and sold in the year 2018 ?

(A) 2:3
(B) 1:2
(C) 1:4
(D) 5:3

46. Find the ratio of the number of units manufactured in the year 2015 to the number of units manufactured in the year 2019 ?

(A) 7:11
(B) 9:14
(C) 7:9
(D) 9:7

47. What is the approximate percentage increase in number of units sold in the year 2019 to the previous year ?

(A) 95%
(B) 75%
(C) 60%
(D) 117%

48. Find the difference between the number of units manufactured and the number of units sold ? (In millions)

(A) 5
(B) 55
(C) 50
(D) 0.5

49. Find the circumference of a circle given the radius = 3cm
(Round to the nearest tenth) Note : ∏ = 3.14

(A) 18.8
(B) 42.7
(C) 17.9
(D) 18.9

50. The average bike sales at speedy bike store across various days are plotted as the box and whisker plot.

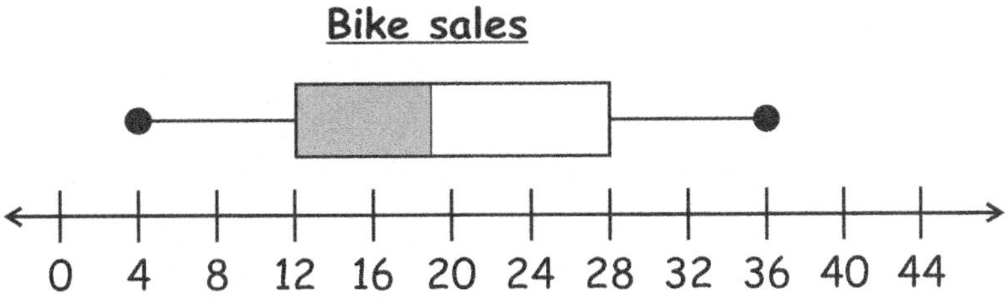

Find the range of the test scores based on the above data

(A) 36

(B) 4

(C) 32

(D) 30

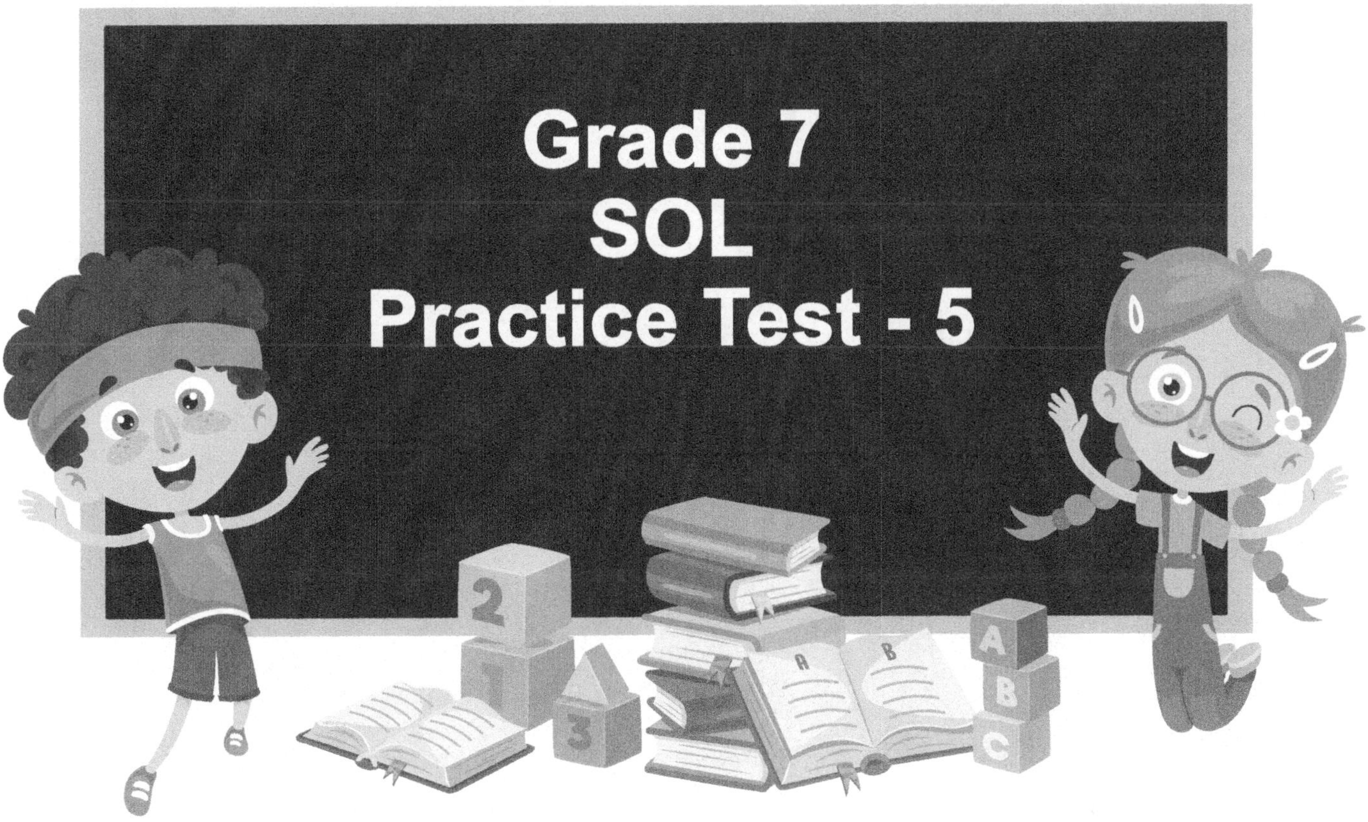
Grade 7
SOL
Practice Test - 5
2
1
3
A
B
A
B
C

1. Simplify

$$3 * (15 + 15) \div 3$$

(A) 30
(B) 27
(C) 45
(D) 15

2. Simplify

$$(-6) - (-7)$$

(A) -9
(B) 1
(C) 4
(D) -11

3. Express the given number in scientific notation

87600

(A) 8.76×10^{4}
(B) 8.76×10^{-3}
(C) 8.76×10^{-4}
(D) 8.76×10^{3}

4. A triangle is 9 in wide and 12 in tall. If it is reduced to a height of 4 in, then how wide will it be ?

(A) 3 in
(B) 2 in
(C) 1 in
(D) 27 in

5. Simplify

$$3*((-30) \div (-10)) - (-14)$$

(A) 23
(B) 31
(C) 38
(D) 8

6. Which of the below is an example of "Rotation 180° about the origin" ?

(A)

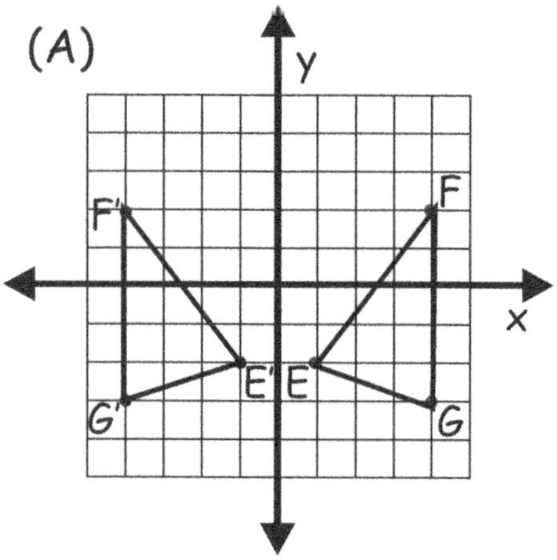

(B)

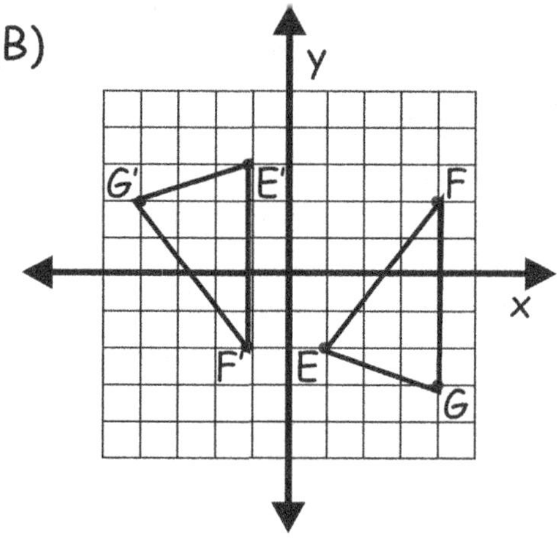

(C)

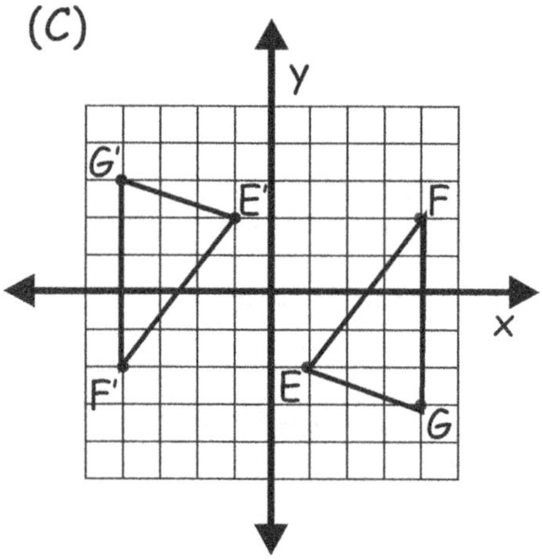

(D)

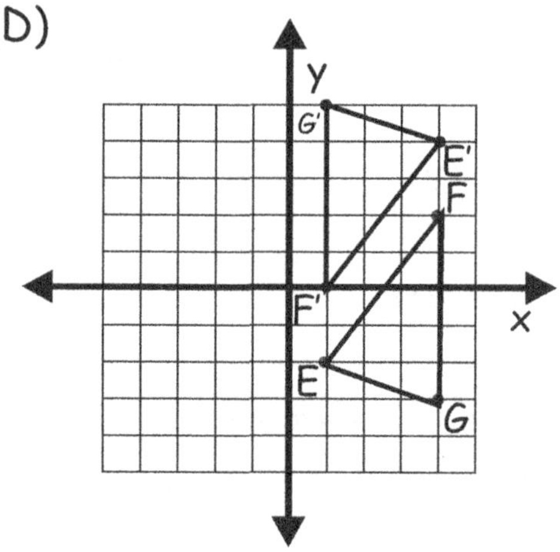

7. Determine the quadrant of the below point belongs to.

(-22, -26)

(A) Quadrant I

(B) Quadrant II

(C) Quadrant III

(D) Quadrant IV

8. Find the volume of the cylinder with a diameter of 14 yd and a height of 5 yd.

(A) 580.6 yd^3

(B) 725.8 yd^3

(C) 631.1 yd^3

(D) 769.7 yd^3

9. Find the area of the given quadrilateral.

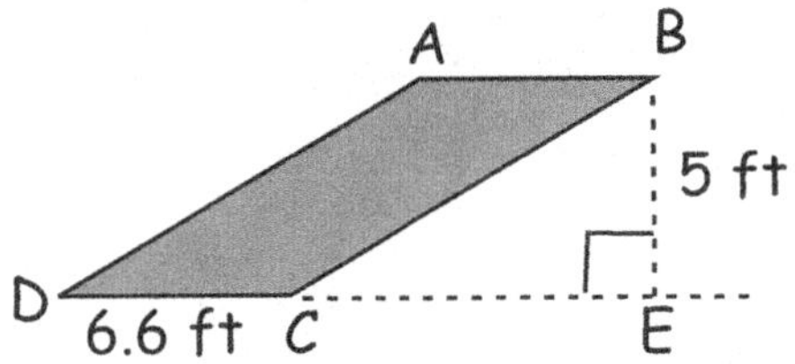

(A) 33 ft^2

(B) 41.5 ft^2

(C) 66 ft^2

(D) 16.5 ft^2

10. Find the total surface area of the given three dimensional shape.

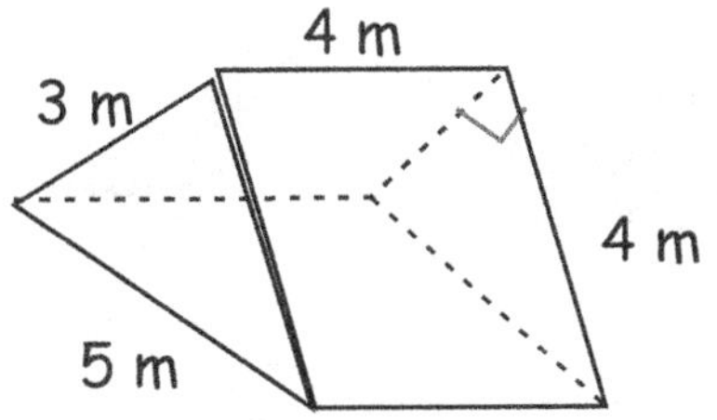

(A) 120 m^2

(B) 49.2 m^2

(C) 60 m^2

(D) 98.4 m^2

11. In how many ways can 3 letters be put into 5 letter boxes when each box can take any number of letters ?

(A) 125

(B) 15

(C) 25

(D) 30

12. Which of the below scatter plots can be plotted using the given data ?

X	Y
2	40
3	80
4	30
5	80
6	10

X	Y
6	20
7	80
7	90
8	10
8	80

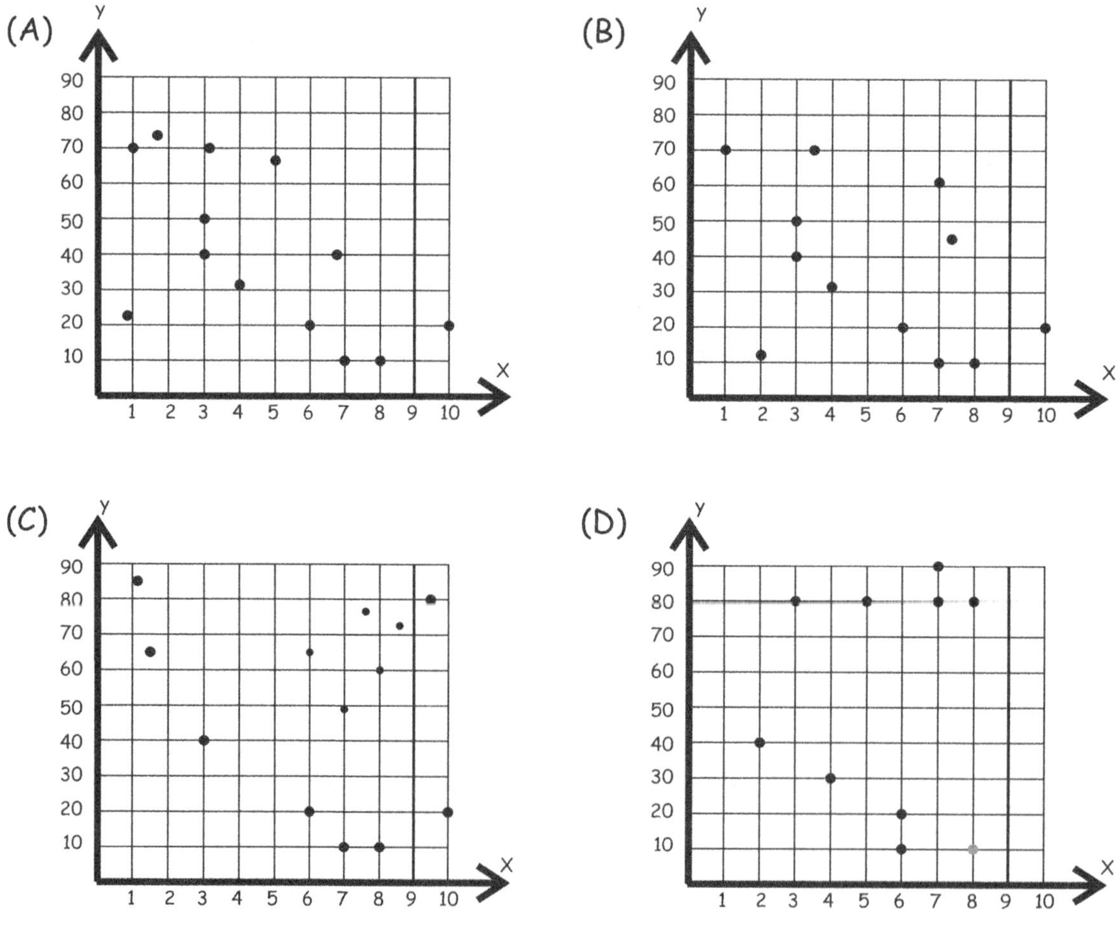

13. Find the median for the given below data.

52 , 49 , 39 , 44 , 47 , 43 , 35 , 44 , 47

(A) 44

(B) 39

(C) 47

(D) 35

14. Find the average for the following data.

Age	11	12	14	16	17	18	20
Frequency	1	2	2	2	6	8	1

(A) 16.88

(B) 15.47

(C) 16.41

(D) 17.16

15. Simplify -14x = -252

(A) -2

(B) 6

(C) 5

(D) 18

16. Translate the verbal expression to an algebraic expression.

n decreased by 9 is 22

(A) 9 + n = 22

(B) $9^3 = 22$

(C) n - 9 = 22

(D) 9 – n

17. Simplify

$$(-11) - (-16)$$

(A) -6 (B) 14

(C) 5 (D) -18

18. Which of the following is the correct statement for the given expression ?

$$n + 9 < 7$$

(A) A number less than 9 is less than 7.
(B) The quotient of 9 and a number is less than 7.
(C) The sum of a number and 9 is less than 7.
(D) Twice 9 is less than 7.

19. Given below is an equation.
What symbol shall be placed in the below box ?

$$7y - 2 \; ? \; 9$$

(A) = (B) +

(C) >= (D) -

20. Translate the verbal expression to an algebraic expression.

x decreased by 14 is equal to 30

(A) $x \cdot 14 \geq 30$ (B) $14 + x = 30$

(C) $\frac{14}{x} = 30$ (D) $x - 14 = 30$

21. Simplify

$$-22 = x + (-15)$$

(A) -37

(B) -7

(C) -2

(D) $1\frac{7}{15}$

22. Which of the below options describes $7a + 21 = 14$?

(A) Equation

(B) Expression

(C) Inequation

(D) Constant

23. Which of the below is the geometric series ?

(A) 10 , 5 , 20 , 40 , 80 , 100

(B) 10, 50, 250, 1250, 6250

(C) $\frac{1}{3}, \frac{2}{3}, \frac{5}{9}, \frac{8}{27}, \frac{16}{81}$

(D) $1, \frac{1}{3}, \frac{45}{9}, \frac{7}{27}, \frac{16}{81}$

24. Macy shared 10 of her muffins with her friends. If she shared $\frac{5}{6}$ of what she had. How many muffins are left ?

(A) 5

(B) 3

(C) 1

(D) 2

25. Olivia is baking cupcakes. The recipe needs $4\frac{3}{4}$ cups of flour. She accidentally pours 8 cups of flour. How many extra cups of flour did she put in ?

(A) $12\frac{3}{4}$

(B) $3\frac{1}{4}$

(C) 8

(D) $\frac{19}{32}$

26. Write an equation showing the relation between x & y.

x	1	3	4	5
y	4	12	16	20

(A) y = x + 3

(B) y = 4x

(C) y = 4 + x

(D) y = x + 12

27. The table below shows Science test scores of Grade 6 student.

46 , 51 , 37 , 45 , 46 , 49 , 42

The teacher missed to add a student score. If the median and mode of the data is same. What is the missing score ?

(A) 46

(B) 51

(C) 49

(D) 42

28. Cathy purchased a fish tank for $149.95. Find the selling price of the fish tank if a discount of 50% is offered on it ?

(A) $224.92 (B) $164.94

(C) $74.09 (D) $74.98

29. Ethan bought a box of tennis balls for $7. How many boxes can Ethan purchase if he had $49 ?

(A) 8 (B) 9

(C) 7 (D) 24

30. The video of the movie Star Wars earned $193,500,000 in rental fees during its first year. Expressed in scientific notation, the number of dollars earned is

(A) 1935×10^{8} (B) 1.935×10^{6}

(C) 1935×10^{5} (D) 1.935×10^{-8}

31. Find the equivalent expression to the below:

2.9(6.06 - 9.95)

(A) 2.9 * 6.06 - 2.9 * 9.95

(B) 2.9 * 6.06 / 2.9 * 9.95

(C) 2.9 * 6.06 * 2.9 * 9.95

(D) 2.9 * 6.06 + 2.9 * 9.95

32. Which of the below is less than 950% ?

(A) 950

(B) $38\frac{2}{5}$

(C) $9\frac{1}{2}$

(D) 95

33. Square ABCD and LMNO are congruent by a scale factor of 3. Then what is the ratio of their areas ?

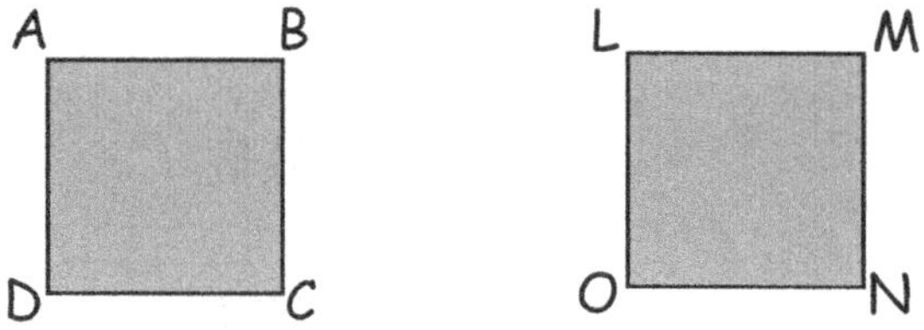

(A) 1 : 9

(B) 2 : 1

(C) 2 : 3

(D) 1 : 3

34. Which of the below is a polygon ?

(A)

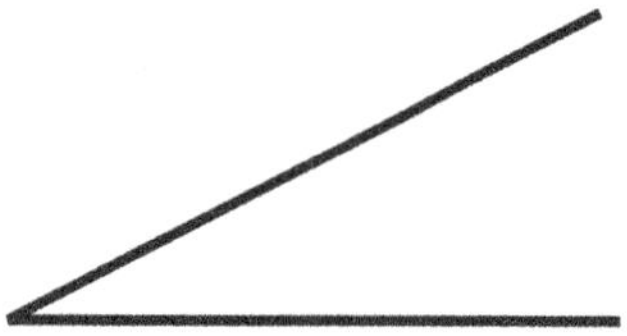

(B)

(C)

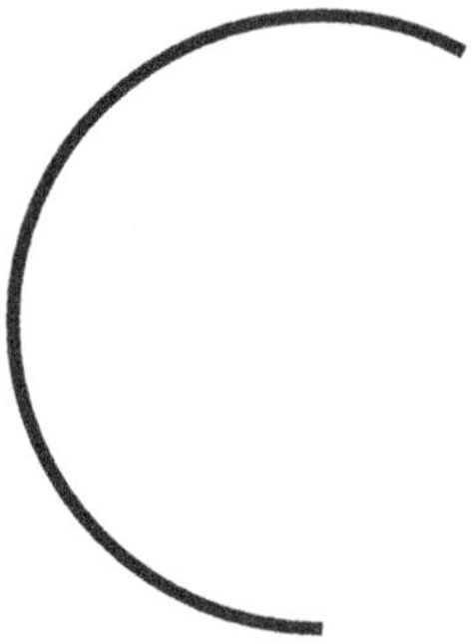

(D)

35. Find the value of p from the below triangles.

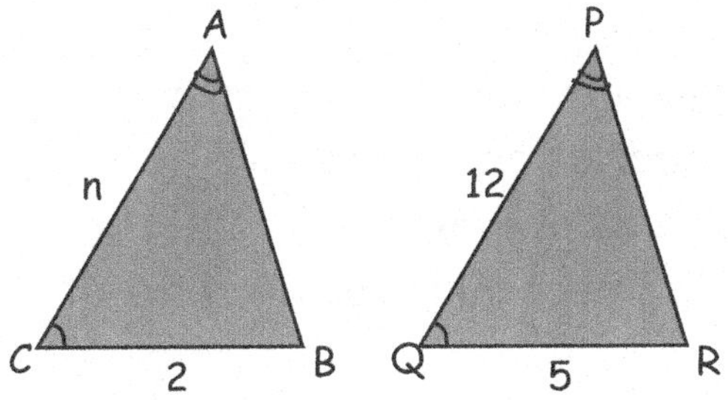

(A) $\frac{2}{n} = \frac{12}{5}$

(B) $\frac{n}{2} = \frac{12}{5}$

(C) $\frac{n}{2} = \frac{5}{12}$

(D) $12n = 10$

36. Temperature at 9 pm is 12° F and the temperature starts decreasing at a steady rate of -1° F per hour. How much is the temperature after 16 hours ?

(A) -4° F

(B) -6° F

(C) 14° F

(D) 40° F

37. Below equation is an example of

$$101 + 0 = 0 + 101$$

(A) Identity Property of Multiplication
(B) Associative Property of Addition
(C) Identity Property of Addition
(D) Commutative Property of Multiplication

38. A local bakery store donates 3% of its sales of every Friday to a local charity. Based on the information given

(A) Total sales amount of last Friday was $950 and donated $28.50
(B) Total sales amount of last Friday was $700 and donated $29
(C) Total sales amount of last Friday was $150 and donated $55
(D) Total sales amount of last Friday was $28.50 and donated $900

39. Which of the below is the property of a kite ?

(A) Two distinct pairs of adjacent sides are congruent
(B) Diagonals of a kite intersect at acute angles
(C) Angles between unequal sides are unequal.
(D) One of the diagonals is the perpendicular bisector of another

40. Which of the below is the property of Rhombus ?

(A) All sides are not congruent
(B) Diagonals are perpendicular bisectors of each other
(C) Every rhombus kite with unequal sides of length
(D) Every rhombus is a not parallelogram

41. Which polygon has one angle more than a hexagon ?

(A) Hexagon (B) Pentagon

(C) Septagon (D) Octagon

42. Express the given fraction as a percentage.

$$\frac{23}{25}$$

(A) 9.2% (B) 152%

(C) 92% (D) 0.92%

43. Express the given fraction as a percentage.

$$\frac{4}{5}$$

(A) 8000% (B) 0.8%

(C) 150% (D) 80%

Perfect printers company spent their income in various categories. The below pie chart represents the expenses.
Answer the questions 26 - 35

Printing Expenses

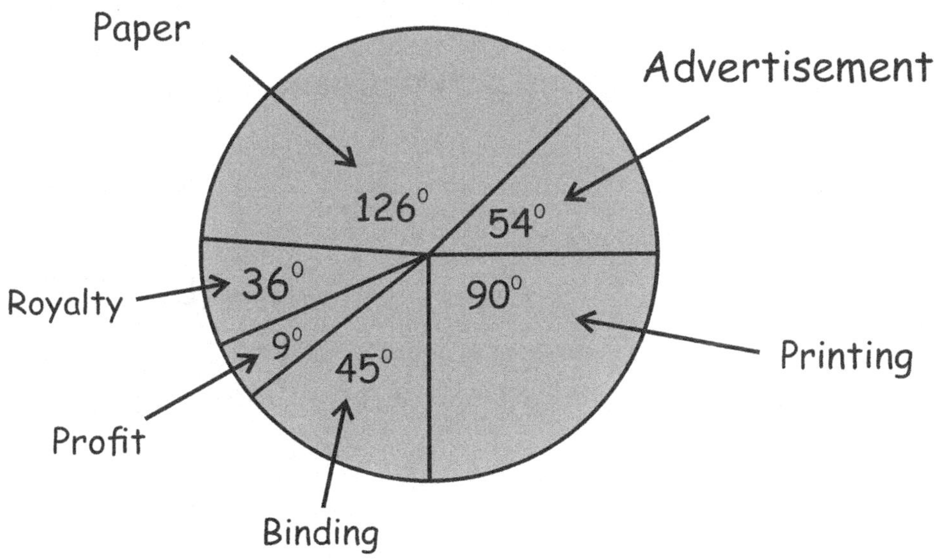

44. Find the ratio of expenses between advertisement and royalty ?

(A) 2:1
(B) 2:3
(C) 3:2
(D) 1:3

45. Find the ratio of expenses between advertisement and printing ?

(A) 5:3
(B) 4:7
(C) 2:5
(D) 3:5

46. Find the ratio of expenses between printing and royalty ?

(A) 5:2 (B) 2:5

(C) 1:2 (D) 2:1

47. Find the ratio of expenses between printing and profit ?

(A) 1:10 (B) 10:11

(C) 11:10 (D) 10:1

48. Find the ratio of expenses between binding and printing ?

(A) 2:1 (B) 1:2

(C) 10:3 (D) 7:5

49. Find the ratio of expenses between printing and paper ?

(A) 7:5 (B) 10:3

(C) 5:7 (D) 3:10

50. Find the ratio of expenses between royalty and profit ?

(A) 4:1 (B) 1:4

(C) 4:7 (D) 7:4

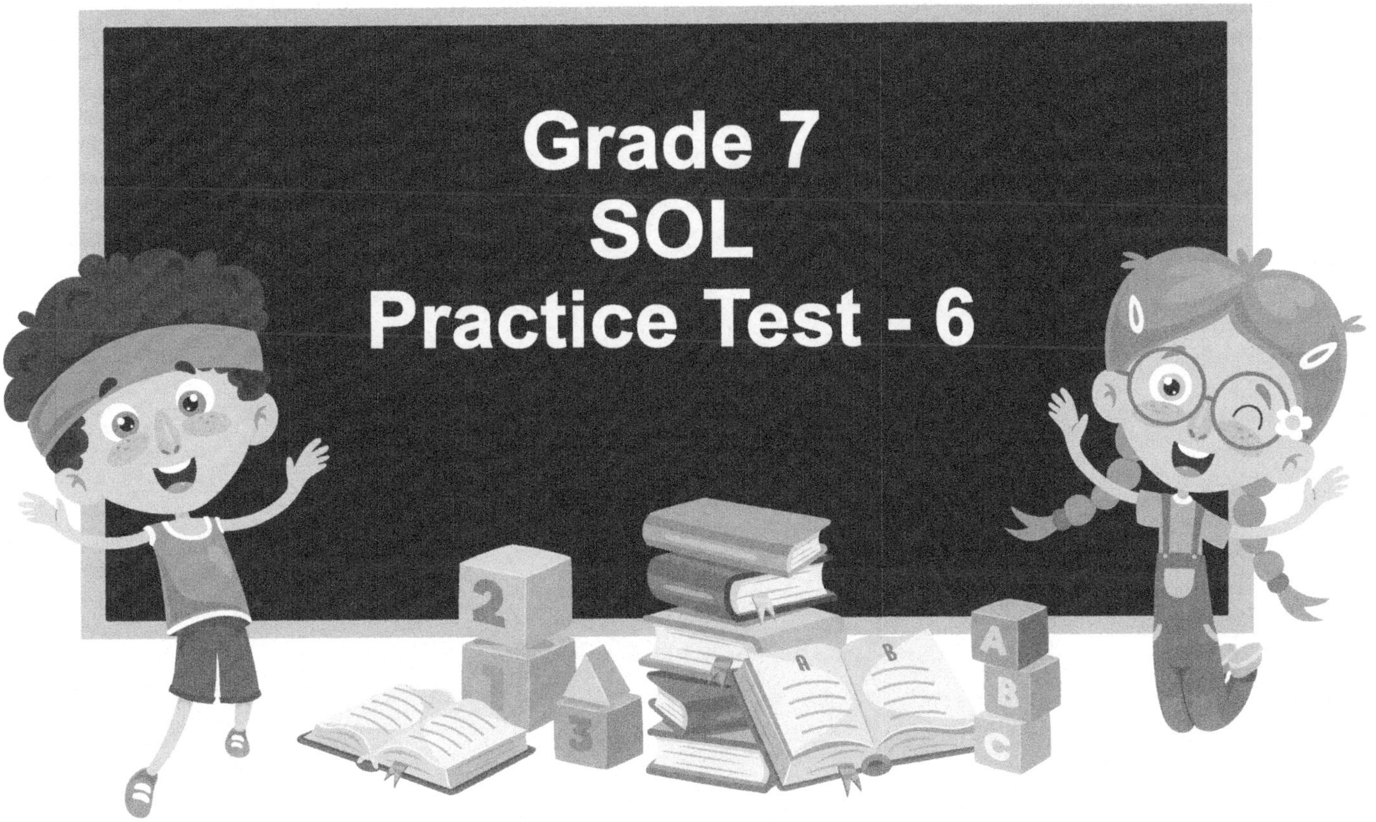
Grade 7
SOL
Practice Test - 6

1. Simplify

$$((-33) \div (-3)) - (4 - 7)$$

(A) 28 (B) 13

(C) 12 (D) 14

2. Simplify

$$(-9) - 2$$

(A) -21 (B) -22

(C) -13 (D) -11

3. Express the given number in scientific notation.

0.0000715

(A) 7.15×10^{-5} (B) 0.715×10^{-3}

(C) 7.15×10^{4} (D) 7.15×10^{-3}

4. 16 honey tangerines cost $12. How many tangerines can you buy for $6 ?

(A) 9 (B) 10

(C) 5 (D) 8

5. Simplify

$$((-20) * 2) \div 2^2$$

(A) -4 (B) -26

(C) -10 (D) -17

6. Which of the below is an example of "Rotation 90° counterclockwise about the origin" ?

(A)

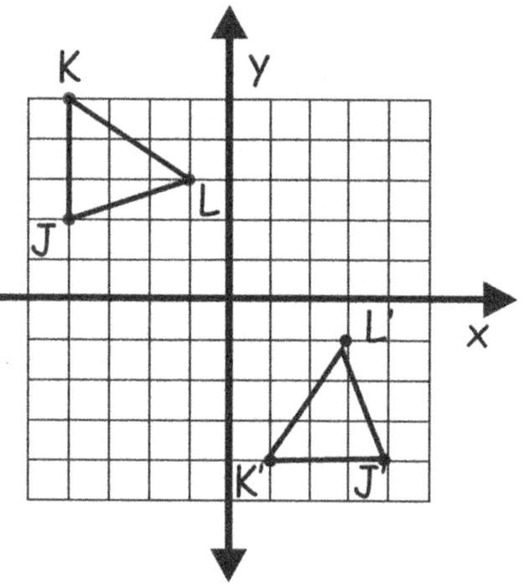

(B)

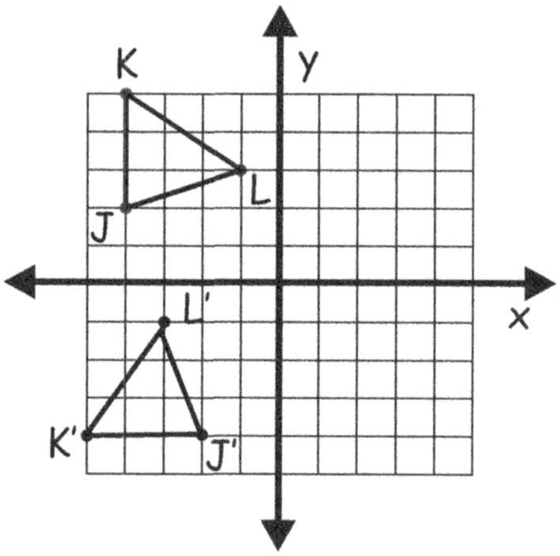

(C)

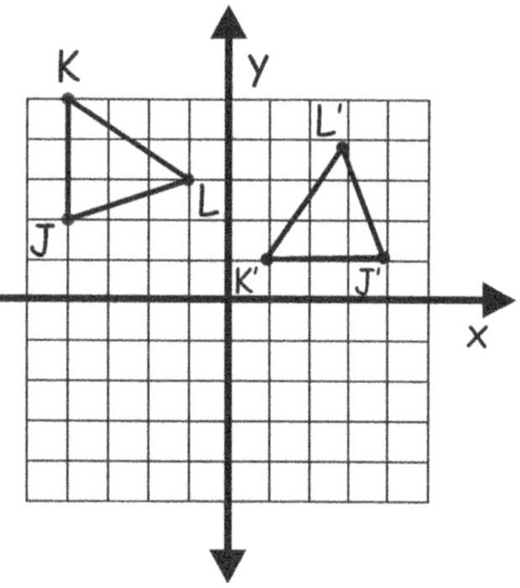

(D)

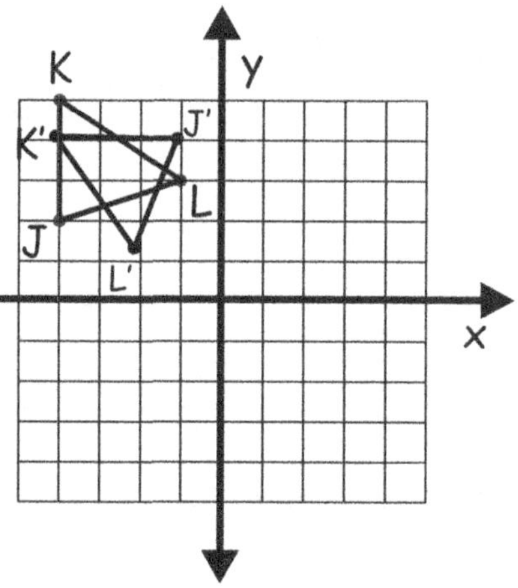

7. Determine the quadrant of the below point belongs to.

(11, -3)

(A) Quadrant I

(B) Quadrant II

(C) Quadrant III

(D) Quadrant IV

8. Find the volume of the cylinder with a radius of 2 yd and a height of 4 yd.

(A) 50.3 yd^3

(B) 61.4 yd^3

(C) 58.8 yd^3

(D) 51.2 yd^3

9. Find the area of the given triangle.

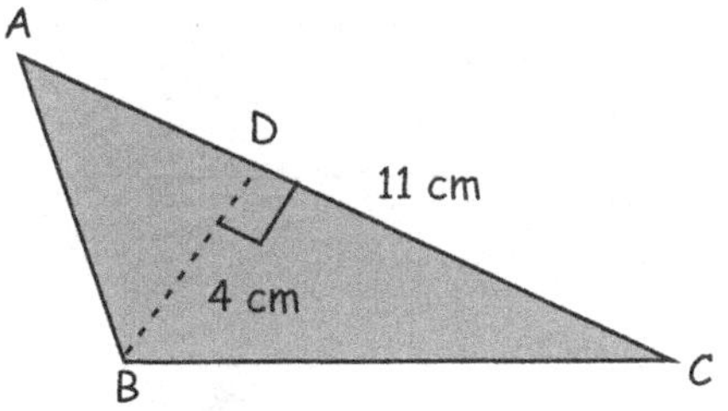

(A) 11 cm^2

(B) 44 cm^2

(C) 22 cm^2

(D) 21.2 cm^2

10. Find the total surface area of the given three dimensional shape.

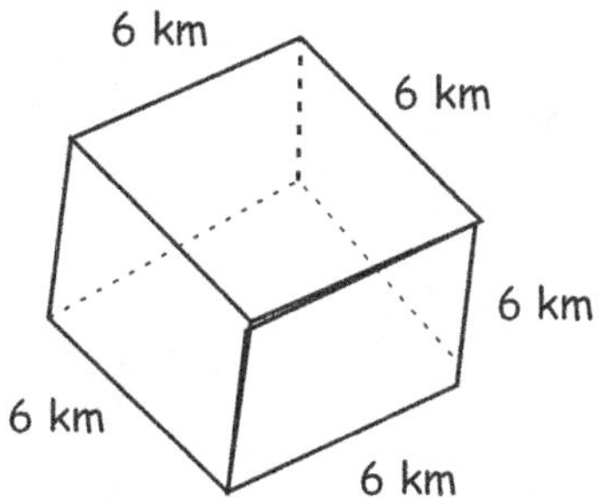

(A) 216 km^2

(B) 30 km^2

(C) 6 km^2

(D) 220 km^2

11. A man has 7 trousers and 10 shirts. How many different outfits can he wear ?

(A) 70

(B) 17

(C) 170

(D) 100

12. Which of the below scatter plots can be plotted using the given data ?

X	Y
1	40
2	55
3	25
4	10

X	Y
5	10
6	20
7	80

X	Y
8	45
9	30
10	30

(A)

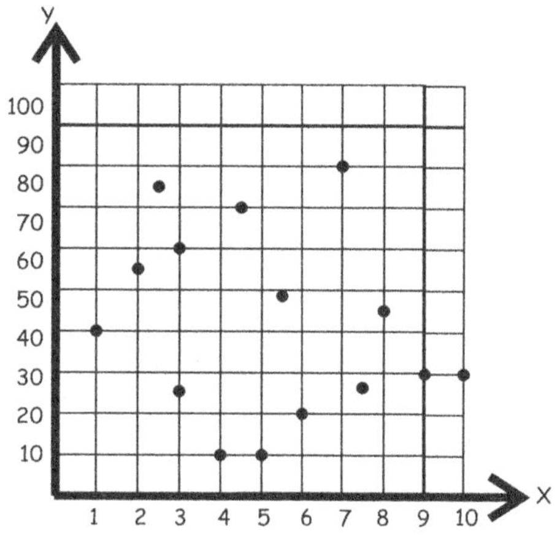

(B)

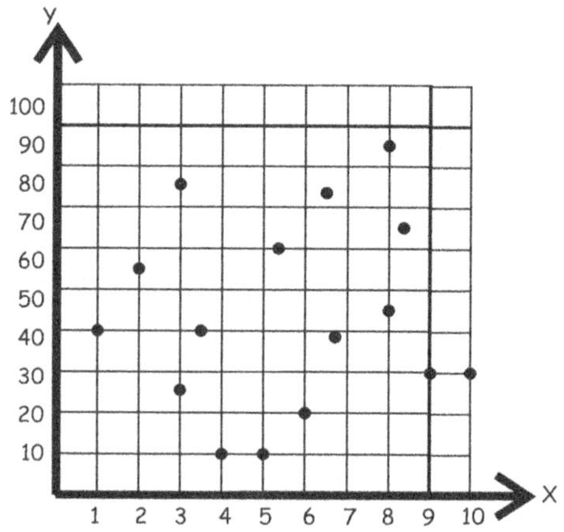

(C)

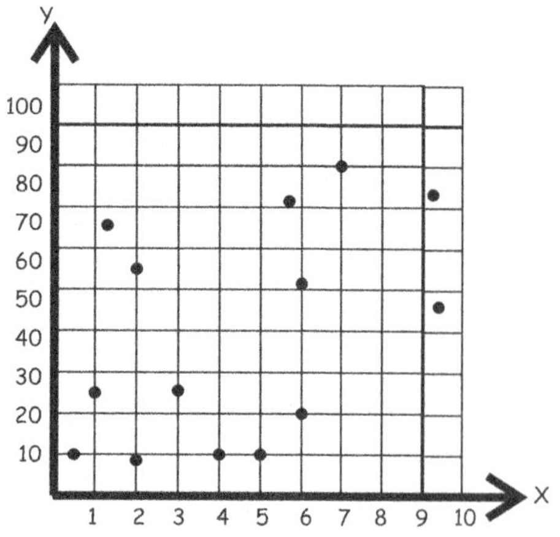

(D)

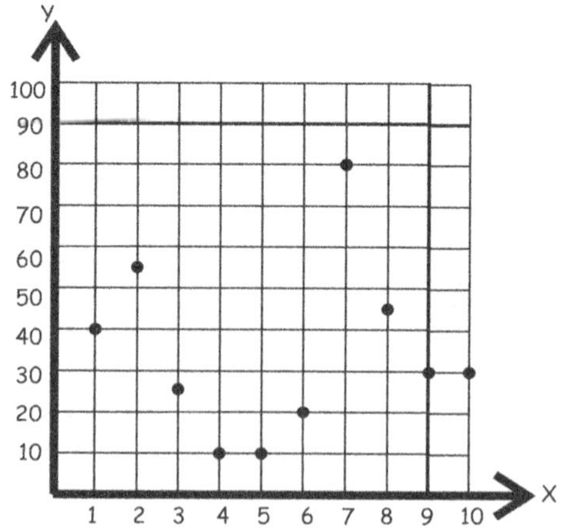

13. Find the median for the given below data.

68 , 66 , 68 , 72 , 71 , 68 , 72 , 68 , 78

(A) 72 (B) 68

(C) 66 (D) 71

14. Find the mean value for the following data.

Hours	6	6.25	6.5	6.75	7	7.25	7.5	7.75
Frequency	1	4	2	5	4	1	3	1

(A) 5.48 (B) 6.76

(C) 6.82 (D) 5.89

15. Simplify $\frac{x}{5} = -19$

(A) $-3\frac{4}{5}$ (B) -14

(C) -95 (D) -24

16. A local store donates 7% of its sales of every Sunday to a local charity. Based on the information given

(A) Total sales amount of last Friday was \$147 and donated \$2100
(B) Total sales amount of last Friday was \$900 and donated \$93
(C) Total sales amount of last Friday was \$2100 and donated \$147
(D) Total sales amount of last Friday was \$300 and donated \$60

17. Translate the verbal expression to an algebraic expression.

The product of x and 11 is greater than or equal to 17.

(A) $x^3 \geq 17$

(B) $x \cdot 11 \geq 17$

(C) $\frac{11}{2} \geq 17$

(D) $x + 11 \geq 17$

18. Simplify

(-5) + (24)

(A) 26

(B) 43

(C) 19

(D) 36

19. Which of the following is the correct statement for the given expression ?

$$\frac{15}{q}$$

(A) The quotient of q and 15.
(B) 15 less than q.
(C) The sum of 15 and q.
(D) The quotient of 15 and q.

20. Given below is an inequality.
What symbol shall be placed in the below box ?

18x + 97 ? 20

(A) <=

(B) +

(C) =

(D) None

21. Translate the verbal expression to an algebraic expression.

The q power of 12 is equal to 8

(A) $12 - q = 8$

(B) $q^{12} = 8$

(C) $12 + q \leq 8$

(D) $12^{q} = 8$

22. Simplify

$$-18 = \frac{k}{4}$$

(A) -22

(B) -72

(C) $-4\frac{1}{2}$

(D) -14

23. Which of the below options describes 6 + 17k ?

(A) Expression

(B) Greater than

(C) Inequation

(D) Equation

24. Which of the below is the geometric series ?

(A) 1, 4, 8, 24, 36, 108

(B) 1, 6, 12, 24, 108, 324

(C) 1, 4, 12, 36, 124, 248

(D) -8, -16, -32, -64, -128

25. Cathy ran 32 miles less than Tony last week. If Cathy ran 17 miles, how many miles did Tony ran ?

(A) 52

(B) 15

(C) 49

(D) 81

26. Violet has $9 to buy a gift for her friend. She has $\frac{3}{10}$ of the cost of the gift. How much more money she needs to save ?

(A) $21

(B) $22

(C) $36

(D) $30

27. Write an equation showing the relation between x & y.

x	1	2	3	5
y	3	12	27	75

(A) $y = 3x + 3$

(B) $y = 3x^2$

(C) $y = x + 2$

(D) $y = 15x$

28. The table below shows number of cars sold across various weeks.

37 , 33 , 37 , 40 , 22 , 41

A weeks cars sales missed in the above. If the median and mode of the data is not altered after adding the missing sales, what could be the missing sales value ?

(A) 41

(B) 33

(C) 22

(D) 37

29. An electronic store in town A sells computers at $850.99. A store is offering a 40% discount today. Find the selling price of the computer ?

(A) $550.20 (B) $590

(C) $510.6 (D) $515.15

30. Tracy has a picture with dimensions of 18 in wide and 6 in tall. She is making a small replica of it with a height of 1 in. What will be the width of the replica she is creating ?

(A) 3 in (B) 5 in

(C) 6 in (D) 108 in

31. According to the 2018 census, the population of New York State was approximately 18,900,000. How is this number expressed in scientific notation ?

(A) 1890×10^{-4} (B) 18.9×10^{4}

(C) 1.89×10^{7} (D) 189×10^{8}

32. Find the equivalent expression to the below:

332.07(112.21 - 121.21)

(A) 332.07 * 112.21 + 332.07 * 121.21

(B) 332.07 * 112.21 / 332.07 * 121.21

(C) 332.07 * 112.21 * 332.07 * 121.21

(D) 332.07 * 112.21 - 332.07 * 121.21

33. Which of the below is less than 612.5% ?

(A) $6\frac{1}{8}$

(B) $612\frac{1}{2}$

(C) $49\frac{4}{5}$

(D) $6\frac{41}{200}$

34. Sides of the rectangles ABCD and XYZW are congruent by a scale factor of 2. Then what is the ratio of length $\overline{CD}$ to length $\overline{WZ}$?

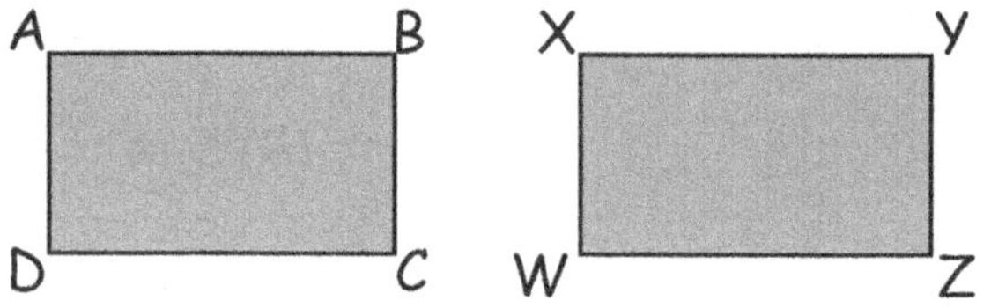

(A) 1 : 1

(B) 2 : 1

(C) 2 : 3

(D) 1 : 4

35. Polygon below is named as ______________

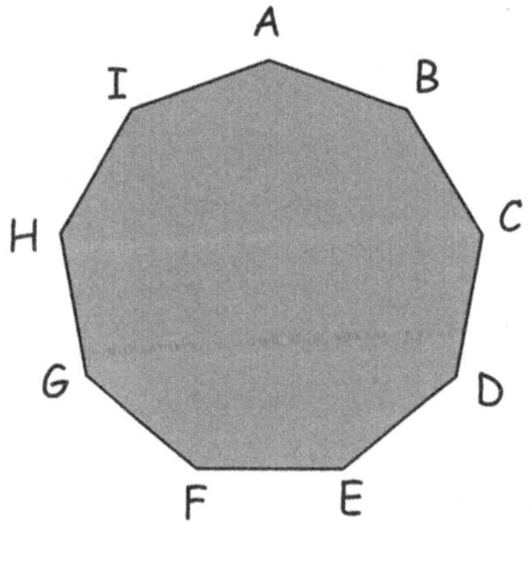

(A) Quadrilateral

(B) Decagon

(C) Octagon

(D) Nanogon

36. Find the value of n from the below triangles.

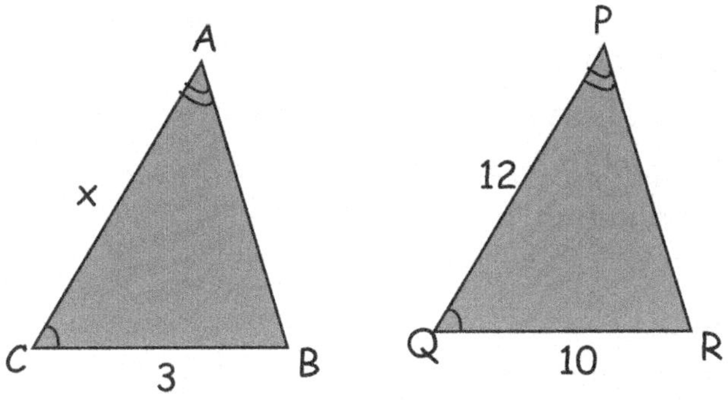

(A) $\frac{x}{3} = \frac{12}{10}$

(B) $\frac{x}{3} = \frac{10}{12}$

(C) $\frac{3}{x} = \frac{12}{10}$

(D) $3x = 120$

37. Zade has a savings of $1080 in her bank last week. She got a bonus of $350 today, she spends $525 on shopping and paid $980 of her student loan. What was her account balance ?

(A) -775

(B) 2935

(C) -75

(D) -425

38. Below equation is an example of

$$470 \cdot (219 + 652) = 470 \cdot 219 + 470 \cdot 652$$

(A) Identity Property of Multiplication
(B) Associative Property of Addition
(C) Distributive Property
(D) Commutative Property of Multiplication

39. Which of the below is the property of a square ?

(A) Two angles in a square are obtuse angles
(B) Diagonals are congruent and bisect each other
(C) Opposite sides are not equal in a rectangle
(D) Every square is a not a rhombus

40. Two six sided dice are rolled.
What is the probability of getting a sum less than 4 ?

(A) $\frac{3}{36}$ (B) $\frac{1}{36}$

(C) $\frac{5}{6}$ (D) $\frac{31}{36}$

41. Which polygon has two less angles than a decagon ?

(A) Hexagon (B) Pentagon

(C) Septagon (D) Octagon

42. Express the given fraction as a percentage.

$$\frac{77}{250}$$

(A) 0.308% (B) 30.8%

(C) 308% (D) 90.8%

43. Express the given fraction as a percentage.

$$\frac{11}{50}$$

(A) 0.22% (B) 2.2%

(C) 11.5% (D) 22%

Based on the below graph answer questions from 44 - 50

44. Find the median of the above data ?

(A) 11 (B) 7

(C) 4 (D) 9.5

45. Find the Mean of the above data ?

(A) 5.67 (B) 8.39

(C) 8.23 (D) 9.88

46. Find the Range of the above data ?

(A) 16 (B) 15

(C) 14 (D) 12

47. What is the Mode of the above data ?

(A) 4 (B) 11

(C) 7 (D) 9

48. Which week has the highest boat sales ?

(A) week 3 (B) week 11

(C) week 1 (D) week 8

49. Which week has the lowest boat sales ?

(A) week 13 (B) week 3

(C) week 10 (D) week 8

50. In which week exactly six boats were sold ?

(A) week 8 (B) week 9

(C) week 2 (D) week 6

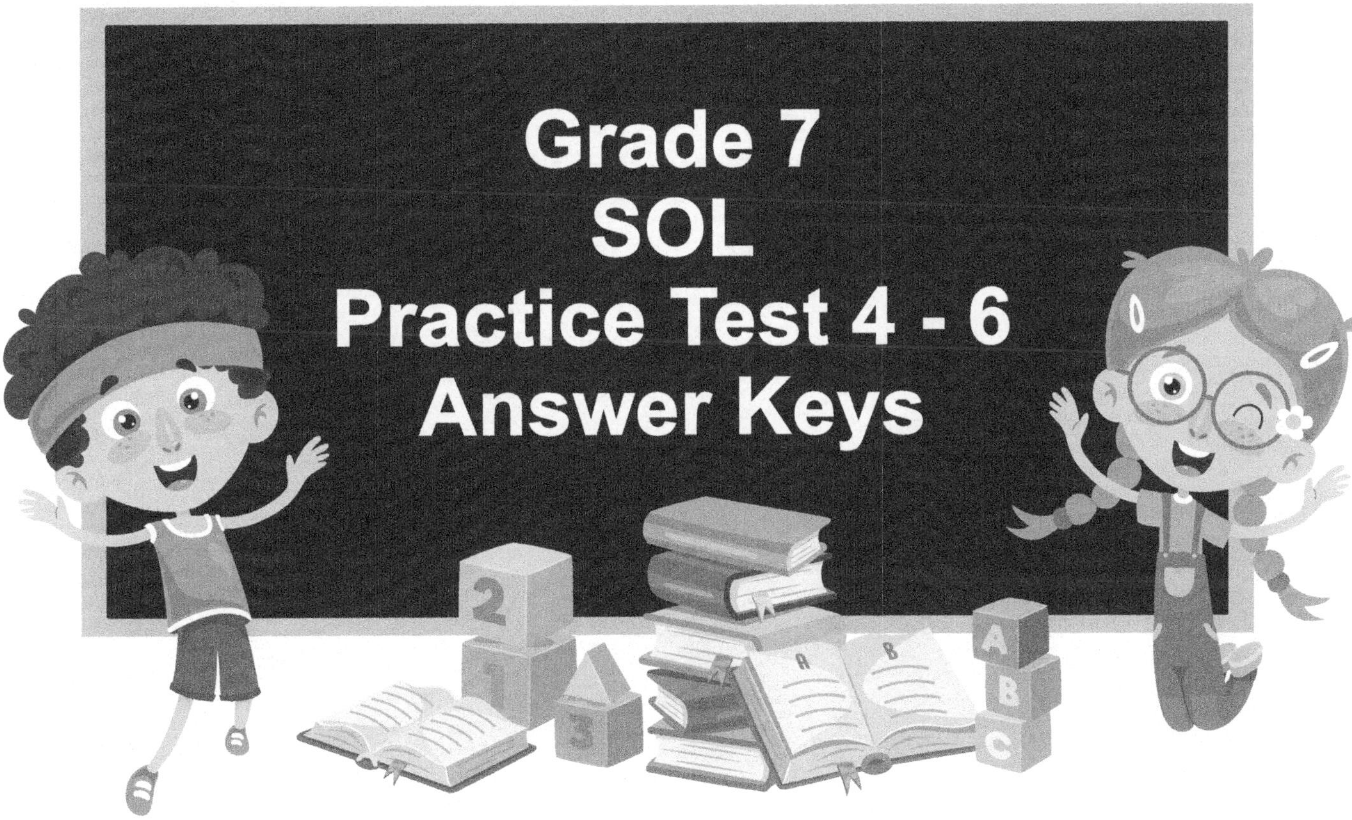
Grade 7
SOL
Practice Test 4 - 6
Answer Keys

Answer Key Test - 4

1. B
2. C
3. D
4. A
5. C
6. A
7. B
8. A
9. D
10. D
11. A
12. D
13. A
14. C
15. D

16. B
17. D
18. B
19. B
20. A
21. C
22. C
23. D
24. B
25. D
26. A
27. A
28. C
29. A
30. C

31. A

32. D

33. B

34. B

35. D

36. A

37. C

38. C

39. B

40. D

41. A

42. D

43. D

44. B

45. C

46. B

47. D

48. A

49. A

50. C

Answer Key Test - 5

1. A
2. B
3. A
4. A
5. A
6. C
7. C
8. D
9. A
10. C
11. A
12. D
13. A
14. C
15. D

16. C

17. C

18. C

19. A

20. D

21. B

22. A

23. B

24. D

25. B

26. B

27. A

28. D

29. C

30. C

31. A

32. C

33. A

34. B

35. B

36. A

37. C

38. A

39. D

40. B

41. C

42. C

43. D

44. C

45. D

46. A

47. D

48. B

49. C

50. A

Answer Key Test - 6

1. D
2. D
3. A
4. D
5. C
6. B
7. D
8. A
9. C
10. A
11. A
12. D
13. B
14. C
15. C

16. C

17. B

18. C

19. D

20. A

21. D

22. B

23. A

24. D

25. C

26. A

27. B

28. D

29. C

30. A

31. C

32. D

33. A

34. B

35. D

36. A

37. C

38. C

39. B

40. A

41. D

42. B

43. D

44. C

45. A

46. D

47. A

48. C

49. A

50. D

GRADE 7
SOL
Score Calculation

Score calculation

If you get this many times correct :	Then your converted scale scor is :
0	000
1	217
2	249
3	268
4	282
5	293
6	303
7	311
8	319
9	325
10	331
11	337
12	343
13	348
14	353
15	357
16	362
17	366
18	371
19	375

20	379
21	383
22	387
23	391
24	395
25	399
26	403
27	407
28	411
29	415
30	419
31	423
32	427
33	432
34	436
35	441
36	445
37	450
38	456
39	461

SCORE

40	467
41	473
42	480
43	487
44	496
45	505
46	516
47	531
48	550
49	582
50	600